MUSCLE CONTROL

by MAXICK

Muscle Control originally published in 1911

MUSCLE
CONTROL

or
Body Development by Will-Power

By

MAXICK

With 54 Full-Page Illustrations
from Special Photographs

PREFACE.

WHEN commissioned by the Publishers to prepare a work upon muscle-control I experienced feelings of great satisfaction, because I felt that the step was being made in the right direction to bring this valuable science prominently before the public.

I do not, and never have, claimed that by muscle-control alone, unaided by mechanical exercises, each muscle may be brought to its highest state of development: but I do claim that mechanical exercise, either with or without apparatus, will never produce the limit of strength and development of which the individual is capable unless combined with muscle control.

The reasons will be fully explained in the following pages; and the proof has been already shown in the many pupils that I have instructed during my three years' stay in England, for muscle-control has a prominent place in my system of physical culture.

A few of the illustrations in the work have been taken from "The Maxick-Saldo System of Physical Culture," by special arrangement with Mr. Monte Saldo and the Publishers, but the great majority of the poses have been specially taken.

The text and photographs have been prepared with the idea of meeting the requirements of the ordinary Physical Culturist, and for that reason all Latin names have been excluded, except where absolutely necessary; simple descriptive terms having been used, so that no anatomical knowledge is required to understand the work.

A chart, showing the principal voluntary muscles of the body, will be found on pages 120 and 121. The reader is earnestly advised to study this, as knowledge of the position of the various muscles will aid him materially to an intelligent accomplishment of the different feats.

MAXICK

ETON HOUSE

ETON AVENUE, NORTH FINCHLEY

THE AUTHOR.

TABLE OF CONTENTS.

MONTE SALDO,
Who has been intimately associated with the Author in this country, and who is
the oldest Englishman to create World's Weight-Lifting Records.

CHAPTER I.

MYSELF.

I TRUST that I shall not be accused of lack of modesty in beginning a book with so seemingly egotistical a chapter-heading. I have not the slightest intention of blowing my own trumpet; but I feel certain that my own personal narrative of how experience gradually revealed to me my method of exercising conscious control over my voluntary muscles will make far more interesting reading to the general public than if I set myself down to the didactic course usually pursued in works of this description.

I foresee that anatomical explanations and references will be unavoidable, but I will endeavor to deal with them as lucidly as possible, urging the reader not to pass such passages over, but to study them, and, for his own sake, to try to fix in his memory the names and positions of the various muscles; because in practicing my methods of muscle-control, one of the most important desiderata is concentration of the mind on the particular muscle to be brought into control.

MY EARLY YEARS.

I am a native of Württemberg, in Germany, and was born on June 28th, 1882.

Being an only child, my father and mother devoted

themselves to my upbringing: every care was bestowed upon me, but I was so sickly an infant, that despite their unremitting attention and the efforts of doctors, the congenital weaknesses developed, and I contracted diseases such as usually spell death to a child of tender years.

Even before I had attained the age of five, I suffered from lung trouble that came to be regarded as chronic, and, eventually, dropsy developed. As may well be imagined, it was generally conceded that I had not long to live. So grave was my condition, that when I had reached my fifth birthday, the official medical man, called in to vaccinate me according to law, refused to do so until he had received a certificate from my family physician exonerating him from all blame in the event of my demise.

Vaccination did not kill me; indeed, I became a little stronger after recovering therefrom, and in a short time began to stand on my legs unsupported for the first time.

ATTACKED BY RICKETS.

But my troubles were by no means over. My parents were of exceptionally small stature, and I was so diminutive for my age, besides being far under normal development, that I could not attend school. And then my general weakness manifested itself in that form of disease most common to sickly children. I became rickety—rickets being a disease attacking the bones. Certain structural anomalies which I bear to the present day will afford ample testimony of the terrible manner in which I was afflicted with the disease.

My parents were in despair, abandoning what little hope they had had of rearing me. The doctors were unanimous that even with the most careful nurturing I should never attain manhood, and that every year of the anticipated short span of life before me would be but one more year of increased suffering.

But, somehow or other, I managed to cheat the doctors, and began to recover a little in health, so that at the age of seven I was able to attend school.

A WEAKLING AMONG THE ROBUST.

And now for the first time it was brought home to me how terrible an affliction is ill health. I had all my life been acquainted with physical suffering; but now I was brought into direct contact with boys of my own age, whose exuberance of spirit was a perpetual source of wonder to me. From wonder I passed to envy of them, and with envy came a sense of humiliation.

I think that if I had not been possessed of a fairly logical mind I should have gone under then. I watched my comrades at play, and was seized with an almost feverish desire to become as strong and healthy as they. But thanks to my temperament, hopeless as my case seemed to be, I never despaired. I was too youthful at the time to devise means to gain the coveted health and strength; but I thought that by imitating my comrades in so far as I could, by eating the same kinds of food they did, I might in time become as they were.

But to do so was not so easy. I begged to be allowed to exercise with weights and dumbbells at home; but my parents did not believe in such things for a sickly child. I was kept on the special diet prescribed for me, and told to abandon all thought even of gentle exercise.

"One as weak as you ought to do nothing but rest as much as possible," was the admonition continually drummed into my ears.

A MOMENTOUS HAPPENING.

I was terribly chagrined, the more so because I felt convinced that my parents were wrong, but there is little doubt that I should have submitted to their ruling, had not an event occurred which had the effect of altering my whole career, and I veritably believe was the saving of my life.

I had reached my tenth birthday, and had improved so far that my health was fairly normal, but I was so undersized and muscularly weak, that I was taken for a boy of only about six to seven years of age. I was therefore at an age to appreciate, with all a boy's interest in such things, the coming of a circus to our little town, especially as the most important item announced was that a strong man would appear, who would, besides performing the usual feats of strength, support twenty-five adult people on a plank.

I cannot describe my eagerness to behold this prodigy of strength: I was nearly heartbroken when my parents refused to take me to the circus. I became so determined to witness the strong man's performance, that I sold all my most cherished belongings—most of them at a huge sacrifice, I am afraid—to my schoolmates, until I had enough money to buy a ticket.

I had no eyes for any other part of the circus, but waited with impatience for the Hercules to appear. This show would not have caused much excitement in these record-breaking days; but I was dumbfounded at his feats, yet was, if possible, even more amazed at the muscular development of the man.

I went home, found a convenient slab of stone, and started in secret to fashion myself a dumbbell. I felt that all that I needed was exercise, and never doubted that in time I should become as strong as the man whose performance I had witnessed. My eagerness to begin overcame my discretion: my father grew suspicious of me, discovered what I was about, and before I had completed it, smashed the crude weight to atoms.

His intention was good, he had been told by the doctors

that I must not exert myself in any way.

And I have no doubt that the smashing of that lump of stone was about the best thing that could have happened, for I was as determined as ever to exercise, and my inventive faculties were now directed to devise means whereby I could carry out my resolve without the use of apparatus which might betray me. And thus I was led to discover that which I have now perfected into my system of

MUSCLE-CONTROL.

During the bedridden days of infancy I had often stretched and contracted my muscles, and it now occurred to me to do this again, but in a more strenuous way, with certain mechanical movements that would tire the muscles.

At first my exercises consisted of weird movements and contortions performed every morning and evening in the seclusion of my bedroom.

And it was then that I began to observe how by certain movements I could contract and relax certain muscles. Assuredly, the best know of all the muscles is the biceps, because it is to that that every boy's attention is drawn, the size of it when contracted being held as a sure indication of a boy's strength and prowess. But in trying to affect this muscle by other movements than by simply bending the arm, my attention was drawn to the way in which other muscles of the arm and forearm responded to these movements.

Now, if at the time I had had some little knowledge of anatomy, of how muscles are usually arranged in pairs which act antagonistically to each other, I should, undoubtedly, have had revealed to me the system of muscle-control which has brought me to my present almost perfect condition of health and strength.

But I knew nothing yet of muscle-relaxation, which, as I shall explain later, is as important to muscle-control as contraction.

MY HEALTH IMPROVES.

But muscular exercises, even as I performed them, had this effect upon me: in about a year's time I had so improved in health and physique that I began to hold my own in tests of strength with my school comrades. And I remember how at the age of fourteen I carried a sack of flour farther than any man in the town had been able to do. Such fame—or notoriety, if you will—did I acquire by this feat, that at the opening of our local athletic club I was invited to become a member, although the age limit of entrants had been fixed at not less than eighteen years.

It is true that I entered as a passive member only, which meant that I was not allowed to handle the heavier weights—for the committee were responsible for accidents to members; but it was an honor, nevertheless, of which I was justly proud, for I felt it to be a significant acknowledgment of the results achieved by my own perseverance.

So encouraged was I that I set to work to bring myself farther along the road of improvement by means of more strenuous exercises than those which had helped me so well hitherto.

I procured light dumbbells; but found that I made no headway: that instead of benefiting, I felt tired after exercise. I reverted to my old system of muscle movements, body contortions, etc.; but although these did not tire me so much, I was conscious that there was little improvement, whereupon I took the dumbbells again in hand, with results as before.

I was nonplussed and chagrined, when it dawned upon me that I was using and tiring my muscles instead of making them stronger. And then it occurred to me that it is not

WORK, BUT NOURISHMENT.

which makes muscles strong. Exercise of the muscles, rational exercise, aids the muscles to obtain nourishment, but as I observed later, rational exercise must be accompanied by mental concentration on the muscles to be exercised

I returned once more to my original method of exercising; but this time I set myself to improve it.

CONTRACTION.

I had already perfected, inasmuch as I could contract every voluntary muscle in my body at will.

But it seemed to me that the more I contracted, the tougher the muscles became, and improvement was checked. Yet, by the aid of a little kneading of the muscles, and by application of the knowledge which I was now gathering from the perusal of scientific works, which, among other things, taught me to use less effort in exercising, I found my development and strength increasing; and this without the regular use of weights or dumbbells.

I was now old enough to leave school, and after some hesitation, derived no doubt from their early fears as to my rearing, my parents decided that I should enter the engineering profession; and I was, accordingly, apprenticed to a local man.

CHAPTER II.

HOW MUSCLE CONTROL WAS REVEALED TO ME.

IN THE engineering shops I had many opportunities of studying the effect of hard labor upon certain muscles.

The smiths and laborers were some of the finest men of natural physique that one could wish to see, and, of course, I expected that they would greatly exceed me in strength. But it was soon acknowledged that, although I was by far the smallest in stature among the youths there employed, I was the strongest.

Yet I was not satisfied; I was puzzled to discover how it was that by means of exercises I had grown strong so rapidly, but that, how, progress to further improvement seemed so slow.

HOW MECHANICAL EXERCISE MAY HINDER MUSCLE DEVELOPMENT.

One day I was watching a journeyman filing metal. I fell to wondering how vaguely why it was that his arm and deltoid development was so small in comparison with that of the rest of his body, knowing, as I did, that the man had worked at the bench for years. Surely, according to accepted theory, it was just these parts which should have been the more developed considering the nature of his work!

I was so interested in this case, that I began to take careful note of other workmen; and my observations at length convinced me that mechanical exercise will not increase bulk or strength beyond a certain degree.

I found out later by experiment that mechanical exercise will only produce good results if interest is directed to the muscles being used. If the mind is directed only to the work being performed, a certain point of muscular resistance is reached; but there it stops. To secure full benefit from the exercise, it is essential that the mind be concentrated on the muscles, and not on the work being performed.

THE CASE OF THE STONEMASON.

Instances by way of example may be given by the hundred. Take the case of a stonemason, who has to use a hammer or mallet for many hours daily, during which time thousands of blows are struck, and the shoulder and arm have to bear the weight, as well as use the mallet.

Now, according to the theories enunciated by many teachers of physical culture, the greater the number of repetitions performed of one exercise, the greater will be the development of the muscles employed. But here is a flat contradiction of these theories, for it will be observed that the majority of stonemasons do not evince anything exceptional in the way of arm and shoulder development.

And the explanation?

Perfectly simple! The stonemason's mind is necessarily concentrated upon the work before him, and he pays little or no heed to his muscles.

Which leads us to another question.

Would the stonemason, or any man wielding a hammer for many hours daily, and concentrating his attention upon the muscles employed, develop colossal muscles in consequence?

The answer is in the negative: for the reason that the mind tires in a short time if concentrated on any one particular object. And then a limit is reached. There would certainly be exceptional development of the muscles upon which the mind was concentrated so long as there was no increase of effort in the blows delivered.

MUSCLE-RELAXATION.

This, as I have already hinted, is the key to proper muscle development. The better to explain how I came to discover this fact, I will return to my narrative.

I had reached the full contraction stage, but had, evidently, come to a standstill.

And why had I come to a standstill? Why did my muscles serve me to a certain extent and then fail me? It was as if they struck work, because I knew that it was not a question of exhaustion of energy, but rather, if I may use such homely language, that the muscles seemed to "stick"—there was a hindrance of free play somewhere.

And there it was that I learned that while one group of muscles is being employed, other muscles are involved which, by their resistance, hinder the free action of the first group. When I had grasped this fact, the idea came to me that, to allow each muscle to put forth to the utmost the energy therein contained, it was absolutely necessary that other muscles must not be allowed to interfere—in a word, they must, by the effort of the will, be relaxed.

And to be able by the exercise of will-power to contract certain muscles while relaxing others antagonistic to them is

WHAT IS MEANT BY MUSCLE CONTROL.

Before proceeding with my narrative, I feel it necessary that I try to explain myself as clearly as possible on this important point, even at the risk of repeating myself.

The newly-born child possesses a certain amount of mechanical control over its muscles, inasmuch as it can move and stretch its limbs in any possible direction: and this is the beginning of the control possessed by the average human being.

According to the art or profession adopted, different groups of muscles are brought more or less under control by the method of constant repetition.

In most cases, the muscles are brought to this state of obedience by external influence, and not by the individual himself.

Many years may therefore be spent in controlling a few groups of muscles that might have been brought under absolute control in a few months, if the muscles had been controlled by the individual in a scientific manner.

The reason that muscles take so long to bring under control by outside influence I have already explained when I pointed out how other muscles are constantly involved, which hinder the movement and control of the muscles particularly required. As time goes on, the unrequired muscles fall gradually into passivity of themselves; but as already mentioned, years may pass before this happens; and possibly the individual may have given up his work in sheer discouragement, having lost hope of ever attaining exceptional or even ordinary skill in his art, profession or craft.

A simple example may be given. Take a student of the piano. However great his musical talent may be, he will never be able to express himself on the keyboard perfectly, until his fingers are under absolute control of his mind. How very few achieve greatness as a pianoforte virtuosi is well known to those interested: for although thousands of students spend their whole time studying at the conservatories, and under eminent masters, the really great may be counted on the fingers of two hands.

This failure in those who evidently possess artistic ability is due always to a lack of proper muscular control. The fingers will not obey the mind of the performer. He knows perfectly well where they ought to go, and where he desires them to go, but they insist upon touching the wrong notes, and in producing the wrong quality of tone.

The trouble is usually, if not always, caused through the actual tendons of the muscles of the hand hindering the action of the flexor and extensors of the forearm.

This brings me once more to the subject of relaxation, which is one of the necessary conditions for successful muscle-control.

Relaxation is just as important as contraction, for unless a muscle be supple enough to lie soft when relaxed, real control is out of the question.

This applies not only to the particular muscle, but to those surrounding, or those muscles which come into direct contact with, and are governed to a certain extent by, the said muscle.

The control of the surrounding muscles will in turn be hindered by the proximity of a muscle or group of muscles that will not absolutely relax.

The toughness of the muscles is known as

MUSCLE-BINDING.

and is usually brought about through indulgence in heavy work, strenuous sport, or incorrect exercising. It is the deadly enemy of agility, or endurance, and is much dreaded by champions in all branches of sport.

That this condition is quite unnecessary may be proved by the fact that the strongest, fastest, and most agile animals have muscles that may be likened in softness to a sponge.

I maintain, therefore, and will show to your satisfaction, that the strongest man, and hardest worker, can retain and add to his strength by educating his muscles and getting them under reasonably good control.

We will take the weight-lifter: the man who is interested in getting heavy weights aloft. He usually starts lifting at the age of eighteen years, and makes rapid strides.

After a year or so, he scarcely appears to improve at all, so slow is his progress: and, usually, by the time he has reached his twenty-fifth year, he seems to have attained his limit. This is not the case with many other sports, followers of which may continue to improve steadily up to thirty-five or even forty years of age; and not only as far as strength is concerned, but in agility, speed and endurance.

This only goes to prove that weight-lifting is the surest way to toughen the muscles and to cause muscle-binding.

CHAPTER III.
I BECOME A CHAMPION.

I HAVE already related how I had come to a full-stop in development: contraction and isolation of the muscles had brought them into prominence, and had nourished them exceedingly; but it needed relaxation to allow them to retain strength and energy, and keep them supple.

I had ere this been able to keep my muscles supple to a certain degree by the aid of massage, but when I had learned of the antagonism of muscles, I strove to find a means whereby I could prevent a certain muscle from operating contrarily to another in action. I thought over the subject day and night, until I found that nature has provided a natural means of massage which is nothing more complicated than the

PASSIVE CONDITION OF RELAXATION.

I worked upon my idea, gradually evolving the system as set forth in the present work, and in a year or two I had brought my body to a most extraordinary condition of development and control, combined with strength that was deemed almost "uncanny." Meanwhile my health became so perfect, that my robustness was just as much the cause of comment as my weakness had been in childhood.

I TAKE UP WEIGHT-LIFTING.

When it had dawned on the club committee how absurd it was to keep me any longer on the passive list, I was re-admitted as an active member, and after six months' practice with weights, I was able to lift with ease in one hand as much as the biggest man there could lift with both hands.

I was asked how it was that I was able to keep my muscles in such a perfect condition of suppleness and control. "The reason," I answered, "is because I first control the muscles, and then lift with these controlled muscles. The ordinary weight-lifter gains his standard by means of weight-lifting exercises with weights, thus his muscles undergo an uninterrupted toughening process, while mine are so supple that I can beat far heavier men than myself at their own pet lifts."

I pointed out one of our champions as an example. He was using and contracting a lot of muscles that could not possibly be of assistance in bringing his weight aloft: with the consequence that he was partially paralyzing and hampering the muscles that, unhindered, could have lifted a much heavier weight.

But I did very little practice with the weights, for I soon recognized that to lift heavy weights and retain true suppleness of muscles was out of the question.

I WIN AN OPEN CHAMPIONSHIP.

It was not long after I had given evidence of my powers as a weight-lifter, that news reached us of an open championship to be held at some distance from our town. I was selected to represent our athletic club: and as there were three classes—heavy, middle, and light weights—and I was not too heavy to enter the light weight, I was entered for all three.

Each club represented sent its own flag bearer. Now our flag bearer was our biggest member, a lusty fellow standing well over six feet in height, and broad in proportion.

When we arrived at our destination, there was much amusement caused by the great contrast in size of our flag bearer and myself. Naturally, the former was taken for the competitor, and we were informed that it was a breach of etiquette that the chosen athlete should carry his club's flag. When the real situation was explained to them, and they learned that it was I who was the competitor, their amusement turned to derisive merriment, which was not to be wondered at, seeing that I weighed barely a hundredweight.

But their merriment turned to wonder when I carried off the light weight; to amazement, when I took the middle weight; and to stupefaction when they saw me returning home triumphant as winner of

ALL THREE CHAMPIONSHIPS.

I was somewhat of a celebrity, and was regarded as a physical phenomenon. But, of course, I knew I was nothing of the kind. I knew instead that I was only an ordinary sort of individual to whom it had been given to discover a rational, natural means to acquiring perfect health, and of employing to the utmost the energy of which a perfectly healthy body is capable.

At the age of twenty-three I went to Munich. I left home because I could not interest myself in the engineering profession. It occurred to me, in an indefinite way at first, that I had a mission in life; and I was well aware that I should not go very far by mere weight-lifting performances and muscular posing before the natives of a place which was little more than a village.

Munich is a famous art center, and there is ample opportunity for a man of good physical proportions to earn a living as artist's of sculptor's model. A visit to the athletic club there brought me to the notice of those who were seeking a model such as myself, and the demand for my services increased so much that I was able to earn a fair amount of money, which was very welcome, as it afforded me the means to pursue my studies in anatomy and physiology.

MY FIRST PUPIL.

I had elaborated my system to such an extent, that I decided to test the efficiency of it on another. I selected for my purpose a man who had been taking a keen interest for years in weight-lifting and muscle development. One day he confessed to me that, try as he would, for the previous year or two he had made no advance either in strength or development.

Here was the very man for whom I was looking! I explained my theories to him, and persuaded him to place himself under my care. He consented, and in a short time he was sensible of improvement, which continued until in about three years' time he had gained a development and control of his muscles almost equal to my own, so much so, indeed, that he succeeded in lifting double his own body weight in a double-handed jerk; a feat which till then had only been accomplished by one man other than myself.

Pupils, posing, lifting and study now occupied me almost exclusively, and I spent a few happy years in this way. Although the lifting kept my development stationary for a time, I worked at my system so steadily that I registered a gradual increase in strength.

Finally I decided to stop lifting altogether, and to go heart and soul into perfecting of my methods.

I COME TO ENGLAND.

But one year later I chanced to see a challenge issued to any middle-weight lifter in the world, and it came from England.

I accepted at once, and came to London; but, as it turned out, the challenge was an old one, and the challenger had, in the meantime, put on so much weight, that he could not get down to the middle-weight limit, and although I was quite willing to contest at catch-weights, the challenger decided that he would not go on with the matter, which was a sore disappointment to me.

I was on the point of returning to Munich when I made the acquaintance of my good friends, Monte Saldo and Apollo.

These gentlemen would not hear of my leaving England until I had put up some records and had given exhibition of muscle-control. What decided me to follow their advice was their declaration that people thought my photographs to be fakes, and my lifts greatly exaggerated.

My blood was up, and in three weeks I was ready, and on the evening of January 19th, 1910, I gave a demonstration before a distinguished assembly of strong men and physical culturists at the Apollo-Saldo School, 9, Great Newport Street.

But I cannot do better than reproduce the account of my demonstration as described by the editor of that decidedly popular weekly, "Health and Strength."

MAXICK'S LIFTING.

*(Reprinted from "*HEALTH AND STRENGTH*")*

Lifters of all ages, weights, and nationalities were there in great force, they having been expressly invited to witness an exhibition by Maxick, of Munich.

Professor Szalay, whose name was associated with weight-lifting more than a decade ago, and who has not inaptly been described as "the father of weight-lifting," was there in all his glory. There were also many of the younger generation, including Messrs. W. P. Caswell (the marvelous ten-stoner), Charlie Kussell (the ten-stone champion of former days), Edward Aston (claimant to the middle-weight lifting championship), W. L. Carquest (the great nine-stoner), Mr. H. C. Tromp van Diggelen, Mr. Reggie Walker (the famous sprint champion), Mr. W. O. Wood (the well-known wrestler), Young Olson, Monte Saldo (Maxick's manager, who acted as stage manager), the Editor of "Health and Strength," and many others.

That Maxick is by way of being a physical phenomenon is beyond question. His muscular control is marvelous. In a series of poses, with which he followed up his lifts, he thrilled the onlookers by the splendor of his development, and the manner in which he "commanded" (that is the word for it) each muscle of his body.

His will seemed to act as commander-in-chief, and at a signal from him, and without any forcing, the latissimus diversi, the abdominals, the deltoids, etc., seemed to do whatever they were told. His body, in fact, was like a transformation scene. One moment he was all chest; the next he was all back; and again you saw his abdominal muscles marshaled, so to speak. It was really very wonderful indeed.

He certainly astonished the onlookers by his weight-lifting feats. He commenced with a number of one-handed lifts, including the 202 lbs. one-handed jerk five times. This seemed quite easy to him. Then one after the other he performed a series of two-handed lifts. The weights were tested by Messrs. Russell, Caswell, Carquest, Szalay, and van Diggelen, who testify that he lifted 222 lbs. clean to the chest, and then pressed it above his head, with his heels together and body erect. His next feat was a 240 lbs. lift clean to the chest. This he pressed above his head in the recognized Continental style.

His next lift of 254 lbs. drew forth a spontaneous encomium from Professor Szalay, who declared that he had never seen such lifting. Maxick raised the barbell clean to his chest, then in a singularly graceful style, pressed it above his head by means of a steady two-handed bent press.

In neither of the three lifts described above did any of the weights come in contact with the lifter's body.

Both the above lifts are claimed as world's records, and it is a pity that we have not as yet a recognized weight-lifter's association, by whom such claims could be officially decided.

Another lift which roused much admiration was the raising of 302 lbs. any way up to the chest. This, which was double his own weight (I should have stated that just before the exhibition he was just under 10 st. 11 lbs.), he then jerked above his head, and really he did not seem to find it very difficult. This was done in the German style: up to the waist, then to the chest, and then aloft.

After the 254 lbs. lift, Maxick made an attempt upon a still further advance upon this. The weight of the bells in this case was kept a secret, only to be revealed in the case of success. Though he made several very creditable efforts, he failed, but it was announced that he would try again on a future occasion.

The exhibition was distinctly interesting, and not by any means devoid of dramatic incident and humor. Nature, when she endowed Maxick with his remarkable physique, threw in with it a very attractive smile. It lit up his countenance every time he made an attempt upon a lift, and it softened into tenderness once or twice when Monte Saldo's pretty, flaxen-haired daughter (aged three) insisted on walking up to him as he was resting and demanded a kiss.

I had always heard that the English were a very conservative race of people, very slow to adopt new theories and ideas, but my experience has proved to be the very reverse, for the English people did not wait for me to prove my theories and assertions on their own physiques, but listened to my arguments and took me on trust, simply because they saw logic and common sense untainted by mystery in my statements.

Assisted by Monte Saldo, I have succeeded in building up a larger clientéle, and our success in curing functional complaints and disorders through the means of muscle-control, combined with suitable mechanical exercises and proper diet, is now a matter of common knowledge.

CHAPTER IV.

WILL-POWER AND MUSCLE-CONTROL.

THE SERIOUS student of muscle-control will soon become aware of the fact that his willpower had become greater, and his mental faculties clearer and capable of increased concentration.

Thus it will be observed that the controlling of the muscles reacts upon the mind and strengthens the mental powers in exactly the same proportion that the control of the muscles strengthens the body and limbs.

Most teachers of physical culture will tell the student to keep his mind concentrated upon the muscles. As the movements are usually mechanical, the advice is necessary, though useless, for monotony tires and jades both body and mind.

The mind is bound to wander during the performance of any exercise that is mechanical, and requires many repetitions.

When, however, an intelligent effort is being made to control a certain muscle, a definite object is being aimed at, and the mind cannot possibly wander. The interest is sustained, and the power of mental concentration gradually but surely developed.

As I have mentioned upon another page, the use of mechanical exercises is necessary for the full development of the whole muscular system, but these may be combined with muscle-control in such a manner that no drudgery of monotony will be apparent.

For lasting and practical results in exercising, it must be pleasurable and energizing; not monotonous and exhausting; and I assert without prejudice to the other many excellent methods of exercising in vogue today, that the greater advances made with the all-powerful march of civilization, the greater will the need of muscle-control become; for a great brain will not be at its best in a debilitated or unfit body, and there will be little time for sports and games, saving for the few. The fight for supremacy will become too keen, and the fit body and quickly-working, responsive brain will be the greatest assets of the bread-winner.

Turning to games of skill, the power of the controlled muscle is undisputed.

Why is it that two men of equally good build, intelligence, keenness, and sight will differ in "form" absolutely?

As an example, take two golfers. They both know exactly where the ball ought to go, but perhaps only one of them can get it in anything like a true direction at every stroke.

One has his driving muscles under control, and the other has not. It may be that the surrounding muscles are hampering or causing a deviation of the muscles required for the particular stroke; but in any case, perfect control and suppleness are not present, or he would make the same stroke in precisely the same manner, and with the same result, as many times as the endurance of the muscles would allow.

The endurance of a controlled muscle is very great indeed.

Firstly, because plenty of blood is available for its use, and secondly, because the bloodflow is unretarded by pressure from the surrounding muscles, for these are all relaxed, and soft also.

The stiff gold student is the despair of the professional instructor. Often one hears the remark: "That fellow will never amount to anything, for he keeps himself stiff, and will not allow the muscles to relax." I agree. He will never amount to anything if he tries to get rid of his stiffness by learning gold or any other game. How can he possibly concentrate his mind on his stroke or game, if he has to think of his muscles as well?

If they had been got into perfect condition by muscle-control, and kept so by a few minutes' daily attention, he would relax automatically, and his whole mind would thus be centered upon his stroke, the correct muscles working unhampered as soon as they were required.

Therefore it must always be born in mind by the student that muscle-control must be regarded in its widest meaning, which is: to relax, restrain, govern, direct and contract the muscles; not only in groups, but singly as far as the connections and adhesions of the other muscles, tendons, and ligaments permit.

A FEW HINTS.

HOW TO GET THE MUSCLES UNDER CONTROL.

Coax the muscles, do not force them.

If undue force be used in an effort to secure quick results, the muscles will toughen, and your object will be defeated.

Just pull or press, as the case may be, following the directions as closely as possible, looking all the time for the particular muscle to appear. As soon as something like the pose has been secured, try to fix your mind upon the exact manner in which you got the result, and relax the muscles, and try to get the pose again in the same way.

Only use a mirror to discover if you have secured the desired control, and not for getting it. Those students who always exercise before a mirror are never confident of getting an exact pose without its aid.

The mirror should be there, placed in such a position that a turn of the head or eyes will show whether the pose has been correctly secured.

If the pose is unsatisfactory, turn the eyes or head away and try again, until you get the muscle or muscles controlled through the feeling, and not the vision.

Before going on to the actual exercises of control, my final advice to the student is:

STUDY ONE EXERCISE AT A TIME AND PRACTICE
PATIENCE.

EXERCISE 1.

RELAXATION.

This is a most important exercise—the beginner must learn to relax all the muscles.

Study the pose (Fig. 1), and it will be seen that not a single muscle suggests contraction.

Think of each part of the body in turn, beginning at the head and working downwards.

Allow each muscle to droop as you think of it; but care must be exercised that, while doing so, you do not contract other muscles which you have already relaxed.

If this exercise is conscientiously performed, you will find that your legs will almost give way under you.

For pose of muscles of the back in this exercise, see Fig. 2. Fig. 2 shows complete relaxation of the muscles of the back. The whole back, as the front, is in repose, and the pose shows how the back should appear in this exercise.

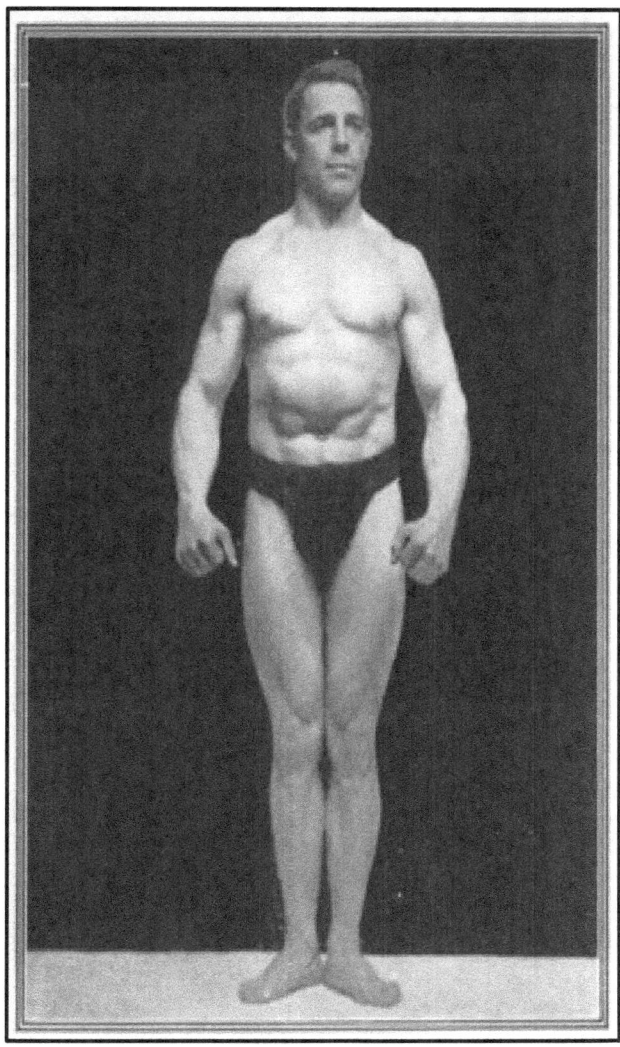

Fig. 1. Complete relaxation, front.

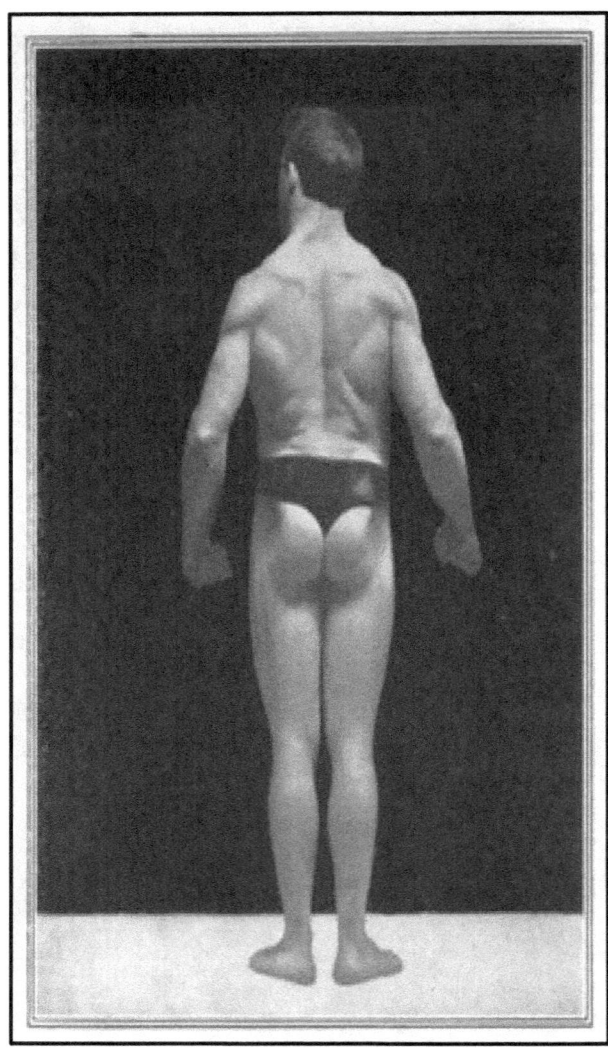

Fig. 2. Complete relaxation, back.

EXERCISE 2.

CONTRACTION.

As soon as you have mastered Exercise 1. and as soon as you feel that the legs are about to give way, strive to contract all the muscles simultaneously.

You will probably find at the first attempts that certain muscles remain in a state of relaxation, not being under proper control of the will.

Relax all the muscles as in Figs. 1 and 2. Then think of each part of the body in turn, from the head downwards, contracting each muscle as you think of it, and retaining each in a state of contraction until every muscle is contracted as shown in Fig. 3.

You will most likely discover when beginning this exercise that you have unconsciously allowed some of the muscles to relax. To these muscles, therefore, you will have to pay most particular attention, contracting and relaxing them until you have them under proper mental control.

These two exercises of relaxation and contraction should be repeated alternately until you are able to accomplish complete relaxation and contraction at will.

Reference to the charts showing the positions of the principal muscles, will be of great assistance to you in thinking of each muscle in turn

The pose, Fig. 4. shows how the back will appear when all the muscles are in a proper state of contraction.

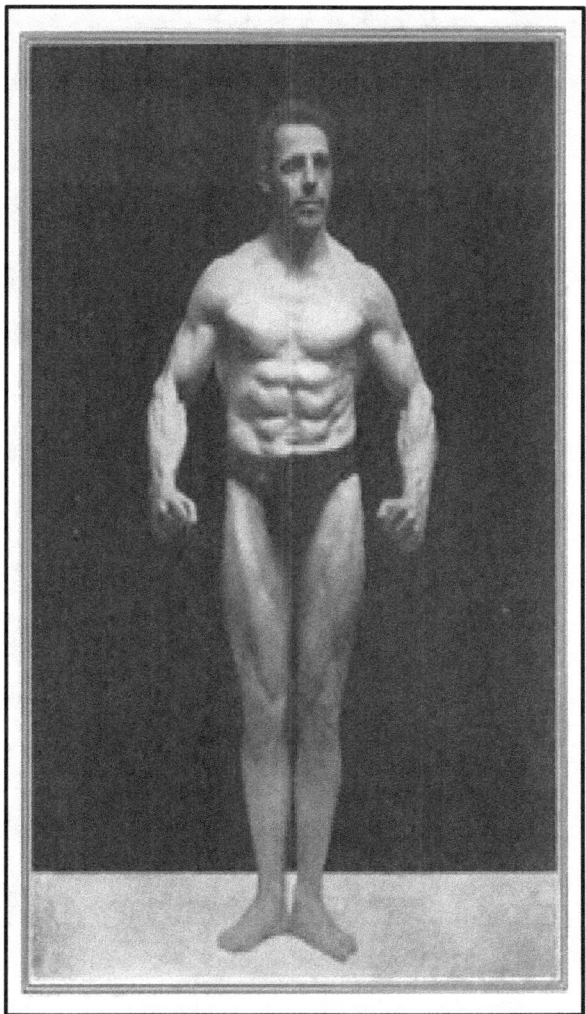

Fig. 3. Total-body contraction, front.

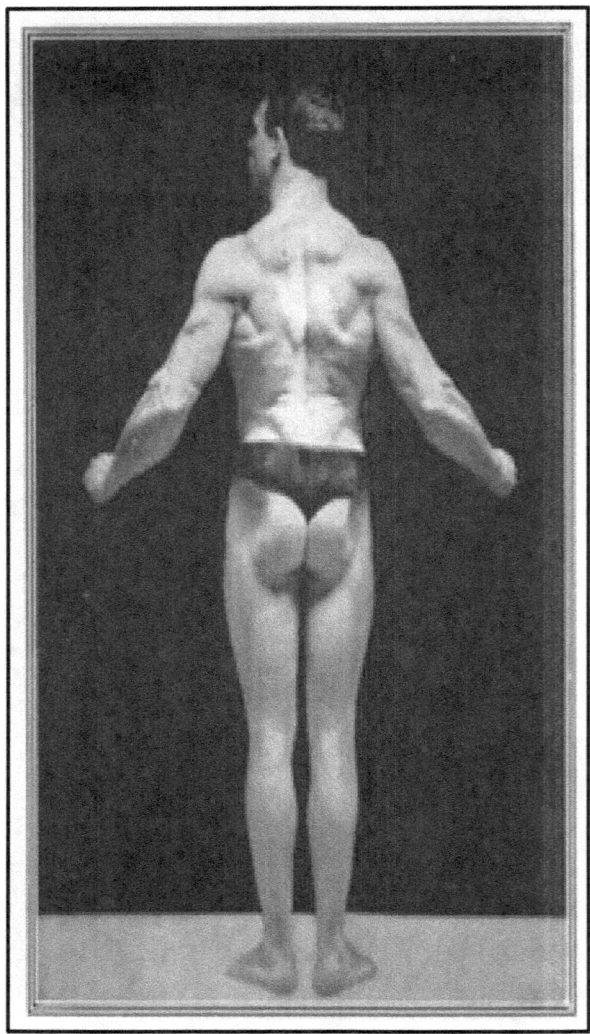

Fig. 4. Total-body contraction, back

EXERCISE 3.

ISOLATION OF THE LATISSIMUS DORSI.

Complete relaxation and contraction having been successfully accomplished, the next step is isolation, which is contraction at will of a particular set of muscles independently of all the other muscles.

We will begin with the big muscles under the arms, the latissimus dorsi.

Pose as in Fig. 5. The hands should be rested lightly upon the hips towards the front, and the muscles allowed to hang limp.

The mind should now be concentrated on the parts indicated by the arrows in Fig. 6. Think of them all the time, and then broaden the back to its uttermost; but without rounding it.

Keeping the back as flat as possible, lift the shoulders, and then drop them when you feel that the back is expanded.

In early attempts, pressure on the hips with the hands, so as to drag down from the shoulders, will be of assistance in broadening the back, though later it will be found that, with practice, these muscles will be readily expanded without such aid.

If difficulty is experienced in accomplishing this exercise without rounding the back, beginners may help themselves by bringing the shoulders forward, rounding the back and pressing with the hands against the waist. Then with the muscles kept expanded, the beginner should strive to bring back the shoulders until the back is flat.

A little practice of this exercise will result in rapid increase of chest circumference.

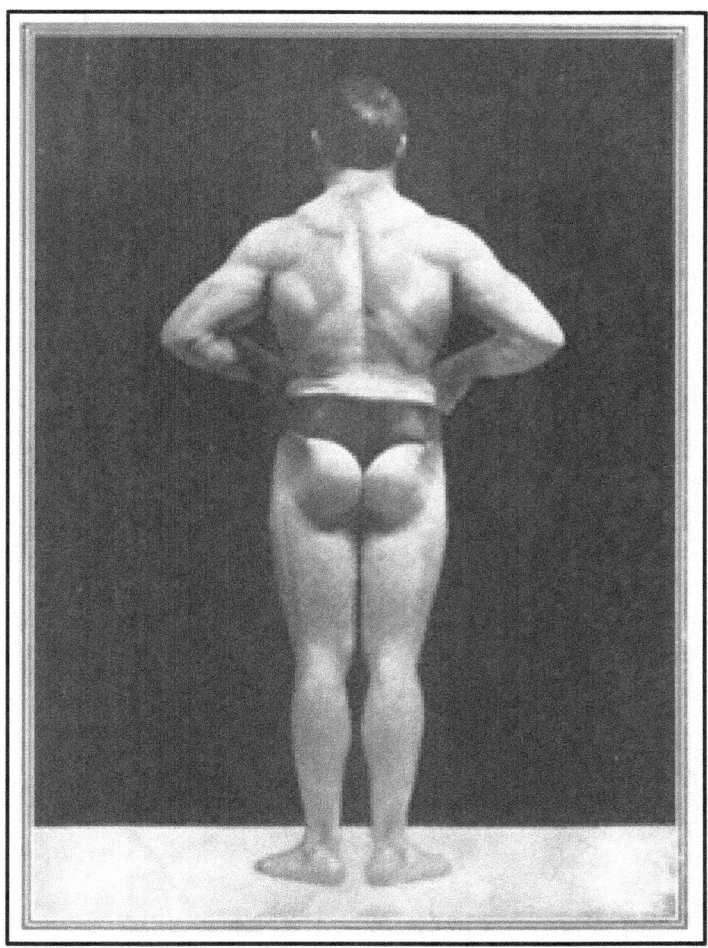

Fig. 5. Latissimus dorsi, relaxed

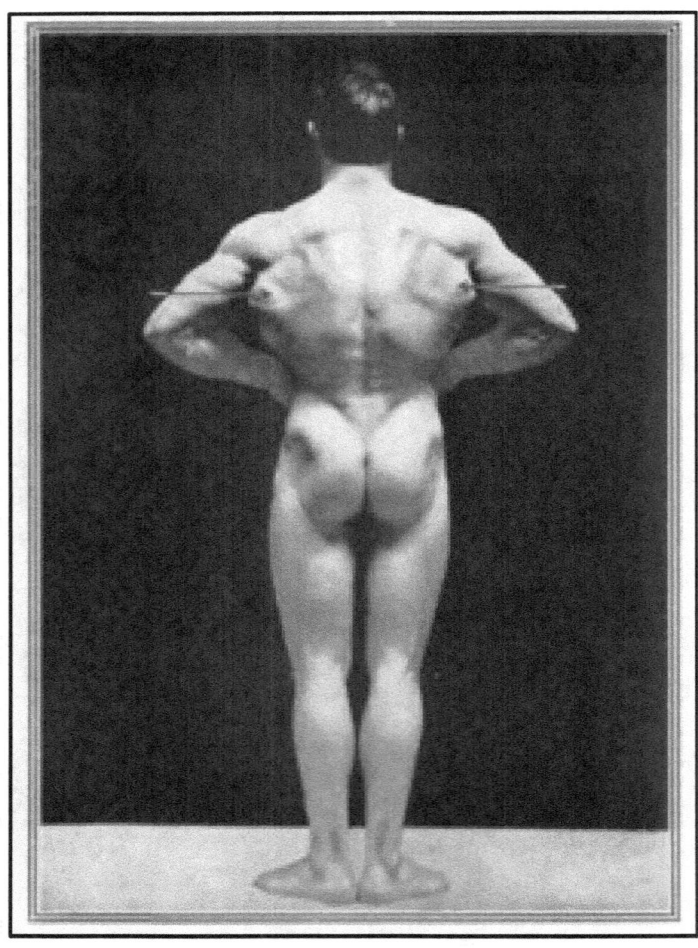

Fig. 6. Latissimus dorsi, contracted

EXERCISE 4.

ISOLATION OF THE TRAPEZIUS MUSCLE.

This shoulder muscle is one of the most difficult to isolate; therefore it has been deemed necessary to explain the method of control by more illustrations than the number devoted to previous exercises.

The pose in Fig. 7 is mainly for the purpose of showing the muscle to be isolated, indicated by an arrow.

SINGLE.

An effort should first be made to isolate the muscle by simple control.

Clasp the hands as in Fig. 8. Drop the right shoulder low, allowing the shoulder-blade to protrude at the back (see arrow-point in Fig. 8).

Now press downwards with the left hand, but resist at the same time with the right hand, keeping the arms almost straight at the elbows.

Practice and experiment until the Trapezius muscle shows in the form of a lump running from shoulder to neck, as in Fig. 7.

Repeat the process with the left shoulder, pressing downwards with the right hand and resisting with the left hand.

DOUBLE.

Having mastered isolation of this muscle on each side singly, an attempt must now be made to isolate them on both sides simultaneously, as in Fig. 9.

If single isolation has been practiced until it can be accomplished with relative ease, little difficulty will be experienced in performing double isolation.

Correct isolation will not have been received until the lines indicated by the arrow-heads in Fig. 9 are clearly defined.

The beginner should understand that although to effect isolation of the muscles mechanical action may be employed at first, it will be found that when the muscular system has by means of these exercises become sufficiently supple and under proper mental control, each muscle will respond with little effort beyond that of mere will-power.

Fig. 7. Trapezius muscle.

Fig. 8. Single isolation of the trapezius.

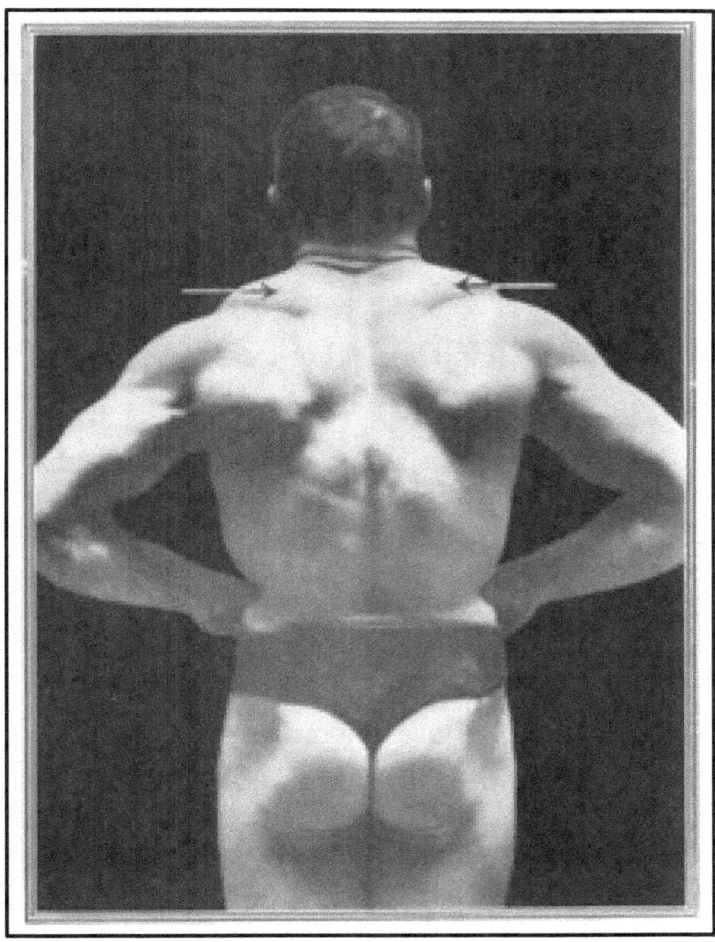

Fig. 9. Double isolation of the trapezius.

Having accomplished double isolation of the trapezius muscle, as shown in Fig. 9, relax the right side only, retaining the left side isolation. Then regain the right side isolation, again securing double isolation.

Then retain contraction on the right side, relaxing the muscle on the left side, then again secure double isolation.

If sufficiently advanced, when you have relaxed the right side, isolate the right hand latissimus dorsi muscle (see Exercise 3), as in Fig. 10.

Then relax the latissimus dorsi muscle, still retaining contraction of the left hand trapezius muscle, and isolate the right hand trapezius, securing double isolation once more in Fig. 9.

Now retain contraction on the right side, relax the left side and isolate the left hand latissimus dorsi, thus reversing Fig. 10.

The latter part of this exercise is a combination of Exercises 3 and 4.

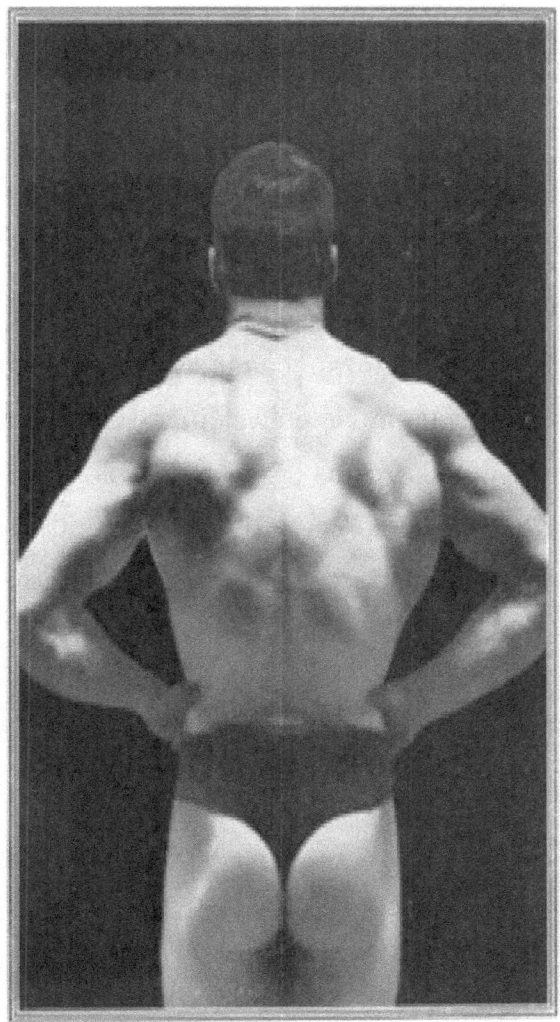

Fig. 10. Back view, right side relaxed.

EXERCISE 5.

CONTROLLED ISOLATION OF THE TRAPEZIUS MUSCLE.

Begin the simultaneous isolation of both trapezius muscles, as shown in Fig. 9. Retaining contraction of them, raise the arms until almost on a level with the shoulders.

Now draw the shoulder-blades together, still retaining isolation of the trapezius muscles, as in Fig. 11.

When the contractions and positions shown in Figs. 7, 8, 9, 10, and 11 have been successfully accomplished, complete control of the trapezius muscles will have been secured.

Mastery of Exercises 4 and 5 will give extraordinary shoulder power and suppleness, with consequent superiority in every sport or occupation in which the arms come into play.

Fig. 11. Isolated trapezius, shoulder blades
drawn together.

EXERCISE 6.

ISOLATION OF THE PECTORALIS MAJOR.

Fig. 12 shows complete relaxation of the great muscles of the chest (pectoralis major).

These muscles are well placed for relaxation, and may be made to dance, if the arm, hanging loosely, be jerked lightly against the body.

But with practice, which means proper control, these muscles may be made to dance without the assistance of the arms, either singly or together.

Clasp the hands as in Fig. 13, then strain the chest as if to bring the arms together, but resisting with the arm muscles at the same time.

Having secured contraction, decrease the pressure of the hands, retaining, as far as lies in your power, the contraction of the chest muscles.

After some practice and concentration of will, isolation of the pectoralis major may be accomplished without the assistance of the hands.

Fig. 12. Pectoralis major, relaxed.

Fig. 13. Pectoralis major, contracted.

ISOLATION AND CONTROL OF THE PECTORALIS MAJOR.

Having mastered the contraction of the pectoralis major without the assistance of pressure by the hands, bring the arms slowly to the horizontal, while still retaining the contraction as already shown in Fig. 13. If successful, the remarkable result shown in Fig. 14 will be secured.

I am not, of course, pressing against anything with my arms to secure this contraction, as the student will discover if he gives to this exercise a reasonable amount of attention.

These chest muscles play an important part in all exercises in which the arms are used, especially in piano playing, as the power imparted by to the arms by these muscles is considerable, especially for inward or downward pressure.

Fig. 14. Pectoralis major, unassisted contraction.

EXERCISE 7.

COMPLETE RELAXATION OF THE ABDOMINAL WALL.

Before any of the exercises of abdominal control can be successfully mastered, complete relaxation of the abdominal muscles must be secured.

A body pose should be sought wherein all strain is removed from the abdominal muscles (Fig. 15).

When there is proper relaxation, the muscle will offer no resistance to the touch. Feel the muscles, and alter the balance of the body until all the muscles are quite soft.

Fig. 15. Relaxed abdominal muscles.

DEPRESSION OF THE ABDOMINAL WALL.

This is effected entirely by external atmospherical pressure; and this exercise is the key to the control, double, and one-sided abdominal isolations.

Deflate the lungs, and then thrust the chest forward (but not upwards), as shown in Fig. 16. If the abdominal muscles are properly relaxed, the atmospheric pressure from without will push them back in the manner shown in Fig. 16, the lungs being empty, and the chest thrust forward.

There must be no abdominal muscular effort to effect this. It is repeated that they must be in a state of complete relaxation, offering no assistance on their own account, and no resistance to the external atmospheric pressure.

If the chest be lifted upwards, the abdominal muscles will not have sufficient play to be pressed inwards.

Fig. 16. Relaxed abdominal muscles in a depressed state.

ISOLATION OF THE ABDOMINAL MUSCLES.

DOUBLE PERPENDICULAR ISOLATION.

Secure the depression as illustrated by Fig. 16, and, without inhaling, raise the arms as shown in Fig. 17. and sway the body slightly backwards and forwards, until the desired contraction has been secured.

It would be useless to lay down any hard and fast rule as to the best position to assume for the accomplishment of this contraction. Many of my pupils have managed to effect it by bending slightly forward.

The contraction should be involuntary, or the whole of the abdominal wall will become involved.

Fig. 17. Double perpendicular abdominal isolation.

CENTRAL SINGLE PERPENDICULAR ISOLATION.

Secure the depression shown in Fig. 16 and without inhaling, press the hands in towards the body, with a slight downward tendency, and the effect shown in Fig. 18 will be secured in a more or less marked degree.

The hands should be placed at the base of the abdominal muscles, one resting in the other for convenience, with the palms upward.

The wrist and part of the forearm may be rested against the pelvis for extra pressure to be obtained; but we have found that the better the position, and the less the pressure, the better the result.

It is a question of correct position and movement. Lean slightly forward when performing this exercise, to give increased play to the abdominal muscles.

For Figs. 18, 19, 20, and 21 we have used photographs of a pupil, Mr. A. W. Beeton of Birmingham, by permission of Mr. J. F. Ritchie, Photographer, of 92, Park Road, Bearwood, Birmingham.

ONE-SIDED PERPENDICULAR ISOLATION.

This is accomplished in precisely the same manner as the central, single perpendicular isolation, as in Fig. 18; but the pressure must now be exerted with one hand, and on one side only, as in Fig. 19.

The simplest way to exercise this contraction is gradually to change the pressure from the center to either side.

The lungs, of course, must be kept deflated all the time.

Fig. 18. Central single perpendicular abdominal isolation.

Fig. 19. One-sided perpendicular abdominal isolation.

CHANGING THE PRESSURE FROM THE CENTER TO THE SIDE.

Fig. 20 illustrates he gradual changing from the center to the sides, which, as already remarked, is the simplest way to secure one-sided perpendicular isolation.

The pressure of the left hand has been partially removed, and the rectus abdominus is, in consequence, giving way beneath the air pressure.

In Fig. 21 we have another remarkable pose by Mr. Beeton, showing that it is possible to secure one-sided isolation of the abdominals unassisted by pressure of the hands, either at the base of the abdomen, or behind the back.

We cannot repeat too often that in all exercises of abdominal control, it must be thoroughly understood that the abdominal muscles are depressed by external atmospheric pressure only, the lungs, being empty, the chest thrust forward, and the abdominal muscles completely relaxed. Any attempt to accomplish this exercise by depression of the muscles by contraction of them is impossible.

The effect on the general health of this exercise is momentous. Practice will relieve all stomachic and intestinal disorders, and will strengthen the abdominal organs, and will operate powerfully against constipation.

Fig. 20. Gradual abdominal pressure
change from side to side.

Fig. 21. Unassisted single perpendicular
abdominal isolation.

EXERCISE 8.

TRUE ABDOMINAL CONTROL.

ABDOMINAL ROLLING.

This is accomplished by securing by contraction a deep depression of any part of the abdominal wall; but it must be borne in mind that all other abdominal muscles must remain absolutely relaxed.

Little difficulty will be experienced to effect this depression, such usually occurring with the unpracticed just below the sternum.

The position of the depression should be gradually changed, working from the apex of the abdominal wall down to the base, and then returning from the base to the apex.

The beginner may assist himself with hands; but it is a course not to be recommended. It is preferable to strive to accomplish this exercise, even at the very first attempts, solely by muscle-control.

Fig. 22 shows the depression about midway between the base and apex of the abdominal wall. Fig. 23 shows the depression almost at the base of the abdominal wall. When correctly carried out, the effect of this abdominal rolling is that of a wave. The beneficial effect of this exercise on the internal abdominal organs will be found to be remarkable.

Fig. 22. Abdominal roll midway between base
and apex.

Fig. 23. Abdominal roll almost at the base of the abdomen.

EXERCISE 9.

ISOLATION OF THE LATISSIMUS DORSI MUSCLES.

WITH ARMS EXTENDED.

Hold out the arms horizontally and in line with the shoulders, making the back as narrow as possible.

Draw the shoulder-blades tightly together.

The shoulders should now present a very narrow appearance.

At this stage, concentrate the mind on the latissimus dorsi muscles, and alternately contract and relax them (see Exercise 3), retaining the position as illustrated by Fig. 24.

Having secured a good "feeling" of the latissimus dorsi, to the exclusion of other muscle groups, by narrowing the shoulders as shown in Fig. 24; now broaden the shoulders to the uttermost, as in Fig. 25.

Compare contraction of the latissimus dorsi in this figure with that illustrated in Fig. 6.

In attempting this feat, it will be found that many other muscles will become involved, including those of the trapezius; but effort should be made to relax all muscles except those of the latissimus dorsi.

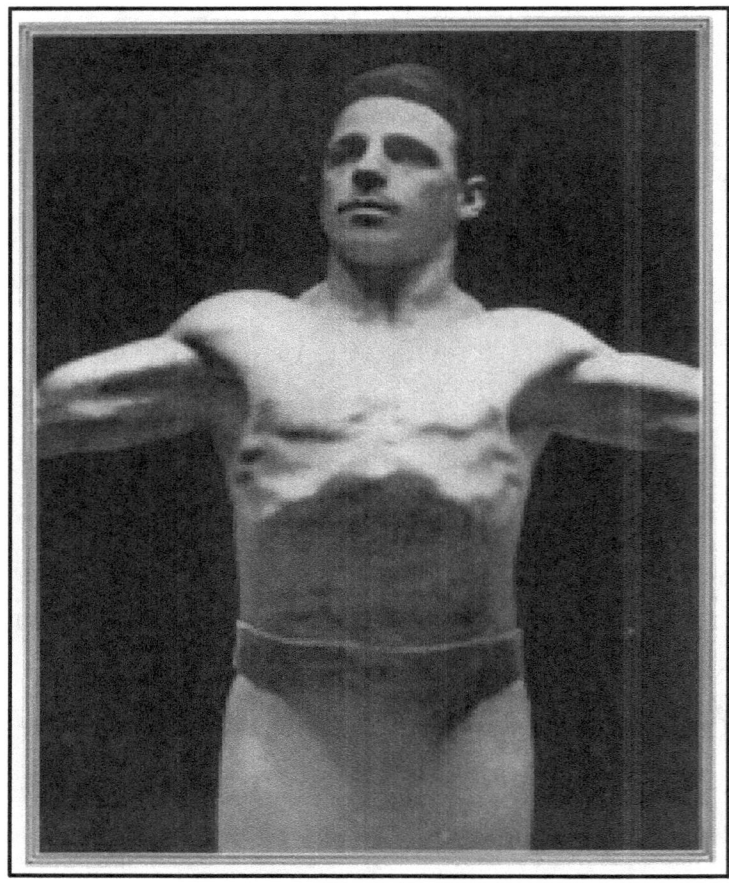

Fig. 24. Beginning pose for isolation of the latissimus dorsi.

Fig. 25. Isolation of the latissimus dorsi with
the arms extended.

EXERCISE 10.

PREPARING THE MUSCLES FOR SHOULDER (DELTOID) CONTROL.

Take up a position as shown in Fig. 26. Hollow the chest, bring the shoulders well forward, and as low as possible.

Take care that all the other muscles are in a state of relaxation and the mind concentrated entirely upon the deltoids.

Having secured a good duplication of Fig. 26, lift the shoulders always to the front as high as it is in your power to do, as illustrated by Fig. 27.

This illustration gives a correct idea of the manner in which the shoulders are to be thrust forward.

Notice in the illustration how the other trunk-muscles are relaxed, hanging, as they are, in folds.

Fig. 26. Chest hollow, shoulders forward and low.

Fig. 27. Shoulders front and high, trunk muscles relaxed.

Having secured the position indicated by Fig. 27, carry the shoulders straight back as far as they will go, without lowering them at all. (See Fig. 28.)

Reach up and back with the shoulders to gain the fullest possible stretch. Endeavor to relax any muscles that seem to hamper the full stretch.

From the position indicated in Fig. 28, lower the shoulders backward and downward until they reach the position shown in Fig. 29.

These four movements are four portions of a circle, and may be combined into one continuous movement if desired.

A fuller explanation is herewith appended.

(1) Shoulders held low and to the front (Fig. 26).

(2) Shoulders lifted high, and to the front (Fig. 27).

(3) Shoulders kept high and carried over to the back (Fig. 28).

(4) Shoulders dropped low and kept over to the back (Fig. 29).

This is a splendid exercise for any sport in which free play and strength of the shoulders is important: i.e., golf, swimming, boxing, fencing, etc.

Fig. 28. Shoulders carried back, remaining high

Fig. 29. Shoulders dropped low and kept back.

EXERCISE 11.

SHOULDER (DELTOID) CONTROL.

Take up a position as in Fig. 30 with elbows pressed against the ribs. Clasp the hands in the manner shown, then alternately pull and push with them, but take care to keep the elbows pressed against the ribs, or the strain will be transferred to the pectoralis major, whereas it is to be concentrated on the deltoid.

In this, as in all other isolation exercises, attention must be given to relaxation of all In this, as in all other isolation exercises, attention must be given to relaxation of all beautiful play of the shoulder muscles.

From position 30, drop the shoulder forward, and allow the shoulder-blade to stand out at the back, as in Fig. 31.

Now push and pull alternately with the hands in exactly the same manner as described for Fig. 30.

Variety may now be had by performing one pushing and pulling movement, in position 30, alternately with one pushing and pulling movement in position 31.

A skillful executant of muscle-control will change these positions almost imperceptibly, the only apparent movement being a rippling of the shoulder muscles.

Fig. 30. Shoulder control initial position, elbows pressed against ribs.

Fig. 31. Shoulder dropped forward, shoulder blade standing out in back.

EXERCISE 12.

TRUE SHOULDER (DELTOID) CONTROL.

Lean forward and slightly to one side (see Fig. 32). Allow the arm to hang limply from the side.

Now drop the shoulders and allow the shoulder-blade to stand out at the back, as in Exercise 4, Fig. 8, bringing the trapezius muscle into play.

Retain this contraction, and lift the arm slightly upward, just sufficiently to contract the shoulder muscles. Keeping the arm at this angle from the body, move it backwards and forwards as far as it will go, describing the portion of a circle with the elbow.

Thus will the shoulder first be contracted by control, and then in turn contracted in all positions.

Practice of shoulder (deltoid) control will be found to be a most effective measure against the attacks of gout, rheumatism, and similar maladies.

Fig. 32. Leaning forward and slightly to one side.

EXERCISE 13.

ISOLATION OF THE SERRATUS MAGNUS MUSCLES.

Interlace the fingers, and then clasp the back of the head as in Fig. 33.

Keep the forearms pressed well against the head. Now bend the neck as far back as possible, looking upwards towards the ceiling.

This will cause the elbows to point upwards as well, which is quite correct, as this will cause the serratus magnus to protrude as seen in Fig. 33.

Now with arms pull the head forwards to the first position, resisting with the neck.

The pull, of course, must be concentrated at the serratus magnus.

Fig. 33. Isolation of the serratus magnus muscles.

EXERCISE 14.

SINGLE ISOLATION OF THE SERRATUS MAGNUS MUSCLES.

Having secured control of the serratus magnus muscles as described in Exercise 13, Fig. 33, drop one arm to the side as shown in Fig. 34.

Do not confine the study of any single isolation to one side only, but practice controlling both sides alternately.

In executing these exercises of control of the serratus magnus muscles, care should be taken that the abdominal muscles are not contracted.

This illustration, Fig. 35, is given to show from the side the manner in which the serratus magnus muscles are controlled.

These exercises are especially important from a health point of view, in that practice is bound to result in greater lung power, as there will be freer mobility of the ribs.

Fig. 34. Single isolation of the serratus magnus muscles from the front.

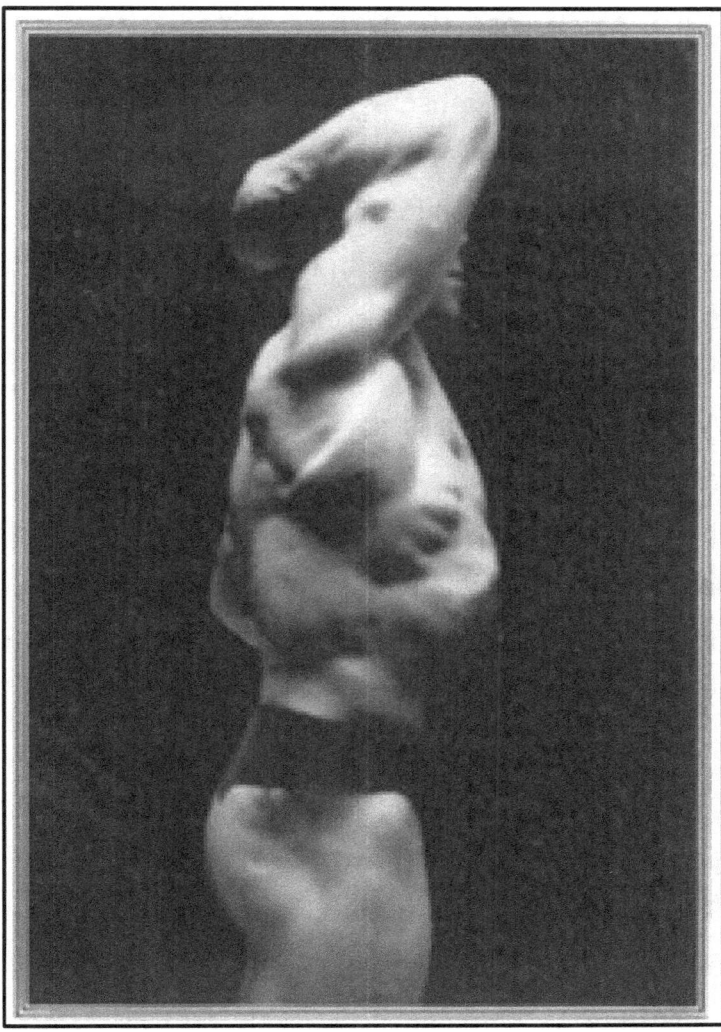

Fig. 35. Single isolation of the serratus magnus
muscles from the side.

EXERCISE 15.

ISOLATION OF THE INTERCOSTAL MUSCLES.

The intercostal muscles are those filling the intervals between the ribs. This exercise is one of the simplest to perform; yet one of the most difficult to describe.

The trouble is usually that either these muscles are lacking in development, or that they are covered with fat. The movement that brings the intercostal muscles into play is scarcely perceptible: for it is secured by leaning slightly to one side, drawing the hip up simultaneously to meet the rib.

If beginners experience difficulty at first in raising the hip, they may help themselves by lifting the corresponding heel from the ground, but effort should be made to dispense with this as soon as possible.

Having drawn up the hip, the intercostals muscles will contract well enough, but they are apt to become bunched up together.

The contraction having been secured, it now remains for the pupil to discover how to spread them out as shown in Fig. 36.

This may be obtained if the body be twisted from above the waist slightly away from the contracted muscles.

The lower part of the body, including the hips, should remain stationary.

Fig. 36. Isolation of inthetercostal muscles.

EXERCISE 16.

LOOSENING OF DELTOID, LATISSIMUS DORSI, & TRAPEZIUS MUSCLES.

Clasp the hands, or interlace the fingers.

Stretch the arms upwards as far as you are able, pulling outwards and sideways, using plenty of energy.

The loosening of the muscles is thus secured by the aid of the shoulder blades.

If the muscles round the chest are tough, a few weeks' practice will be necessary before the result shown in Fig. 37 will be secured.

In the pose illustrated, the deltoids are doing the work, and the latissimus dorsi are being made supple by stretching.

This exercise has the additional advantage of giving suppleness to the shoulders.

Having secured the effect shown in Fig. 37, bring the hands down on to the head, still pulling outwards and retaining the expansion of the shoulder blades, as in Fig. 38.

Do not draw the shoulder blades together, but broaden the back to the uttermost.

Depend entirely upon a correct outward pull for the accomplishment of the effect illustrated.

This exercise will ensure symmetry of form, for if the effort be not even on both sides, it cannot be correctly accomplished.

Fig. 39 illustrates exactly how the latissimus dorsi muscles should appear from the front, when the position depicted in Fig. 38 has been secured.

Fig. 37. Loosening of the shoulder blades.

Fig. 38. Loosening the muscles, from the back.

Fig. 39. Loosening the muscles, from the front.

EXERCISE 17.

CONTROLLING OF DELTOID, LATISSIMUS DORSI, AND TRAPEZIUS MUSCLES.

Secure the contraction illustrated in Fig. 38.

Relax the power from the outward pull, and push the hands strongly together.

The shoulder blades have dropped right hack to normal position, and a different portion of the deltoid (shoulder) is brought into action. (See Fig. 40.)

Retain the contraction of the back muscles, as illustrated by Fig. 40, release the pressure of the hands, and remove them from the head, flexing the biceps as shown in Fig. 41.

Concentrate the mind in quick succession upon biceps, deltoid, trapezius, and latissimus dorsi muscles, retaining control (in contraction) of all four set simultaneously, then securing them separately.

Once control of these muscles has been secured, in the manner indicated, there will ensue a big increase in weight-lifting capacity.

Fig. 40. Shoulder blades dropped, hands pushing strongly together.

Fig. 41. Hands removed from the head, biceps flexing.

EXERCISE 18.

CONTROL OF EXTENSOR MUSCLES OF THE ARMS.

From position of repose as shown in Fig. 42, gradually contract, while thinking of them, all the extensor muscles of the arms.

It will be observed that in the illustration my hand is clenched; but contraction of the extensor muscles must be performed without the assistance of clenching the fists. The beginner had better attempt this contraction with the hands held loosely open.

Now lock the arm at the elbow, by contraction (and pushing with) the triceps. (See Fig. 43.)

Retaining the locked elbow, push the arm backwards as far as it will go.

Return to the original position (Fig. 43), and then bring the arm slowly up in front of the body, right to full stretch above head.

Throughout the movement concentrate the mind upon keeping the arm absolutely locked, by pushing with the triceps.

Fig. 42. Extensor muscles relaxed

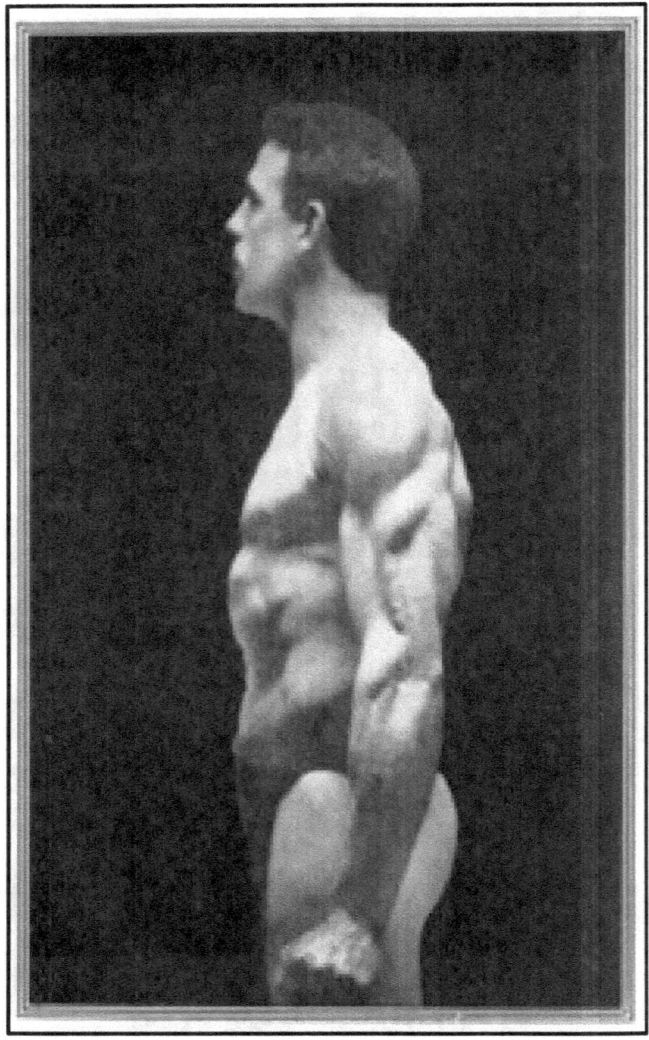

Fig. 43. Contraction of the triceps.

EXERCISE 19.

CONTROL OF THE FLEXOR MUSCLES OF THE ARM (BICEPS).

Bend the body forward, holding the elbow almost at right angles to the body.

Contract the biceps as shown in Fig. 44.

Note well that the shoulder is dropped low and the upper arm and elbow pushed well away from the body.

This position secures a magnificently full and powerful contraction of the biceps without much effort.

The elbow may be lifted even further from the body, and the shoulders dropped still lower.

This position will enable the student with a well developed arm to make the forearm and biceps meet.

In Fig. 45, a similar position of the arm is shown as in Fig. 44, but the biceps is shown full instead of in profile.

The student's attention is called to the fact that the fist is clenched.

To secure full contraction of the biceps, it is absolutely unnecessary to clench the fist, as will be shown in Fig. 46.

The contraction of the biceps is still retained with the open hand and a slack waist. (See Fig. 46.)

I know that beginners will, almost instinctively, clench the fist, when contracting the biceps; the student is, therefore, advised to attempt this exercise always with an open hand.

Having secured position shown in Fig. 46, turn the palm of the hand outwards, still retaining the contraction of the biceps. (See Fig. 47.)

The biceps will have lengthened somewhat, but it is possible to keep it contracted and hard throughout.

Return the forearm to Fig. 46, clench the fist as shown in Fig. 45, relax and repeat.

The four contractions shown in Figs. 44, 45, 46 and 47 will incidentally bring the forearm under control, if the attention be transferred from the upper arm to the forearm.

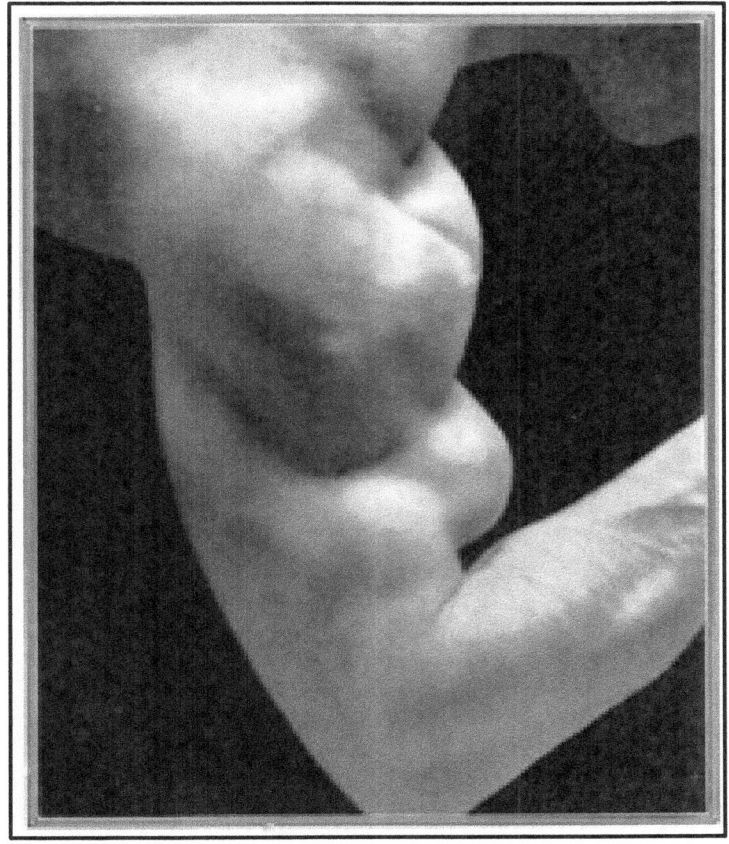

Fig. 44. Contracted biceps in profile.

Fig. 45. Contracted biceps in full.

Fig. 46. Contracted with open hand.

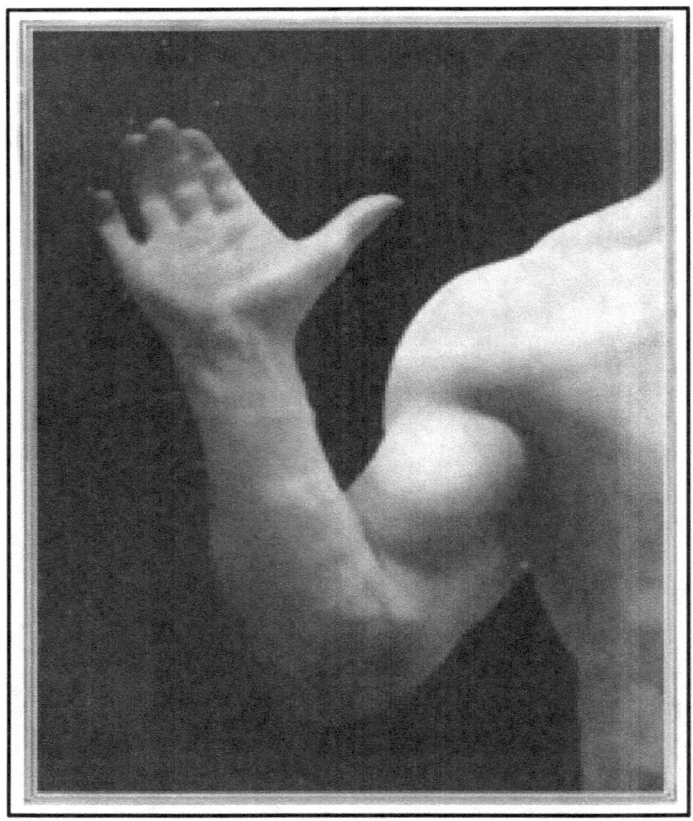

Fig. 47. Contracted with palm out.

EXERCISE 20.

CONTROL OF THE EXTENSOR MUSCLES OF THE THIGH.

Stand upon a rough surface where the feet will not slip, and strain as if to force the legs apart, but keeping the knees straight.

The effect shown in Fig. 48 will be reproduced.

Considerable concentration will have to be used to prevent the knees from bending sideways.

The stiffer you are able to keep the knees, the greater will be the effect upon the thighs.

Place one foot a few inches to the front.

Rest the foot softly upon the ground; do not press or put any weight upon it.

Now concentrate the mind upon bringing the extensor muscles into high relief by contraction alone. (See Fig. 49.)

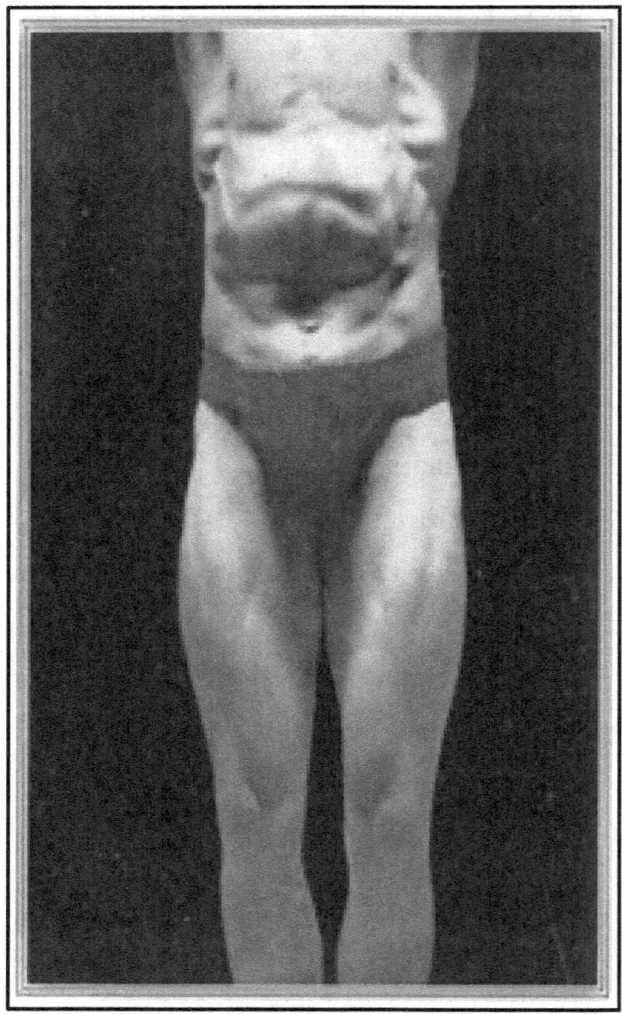

Fig. 48. Strain as if forcing the legs apart

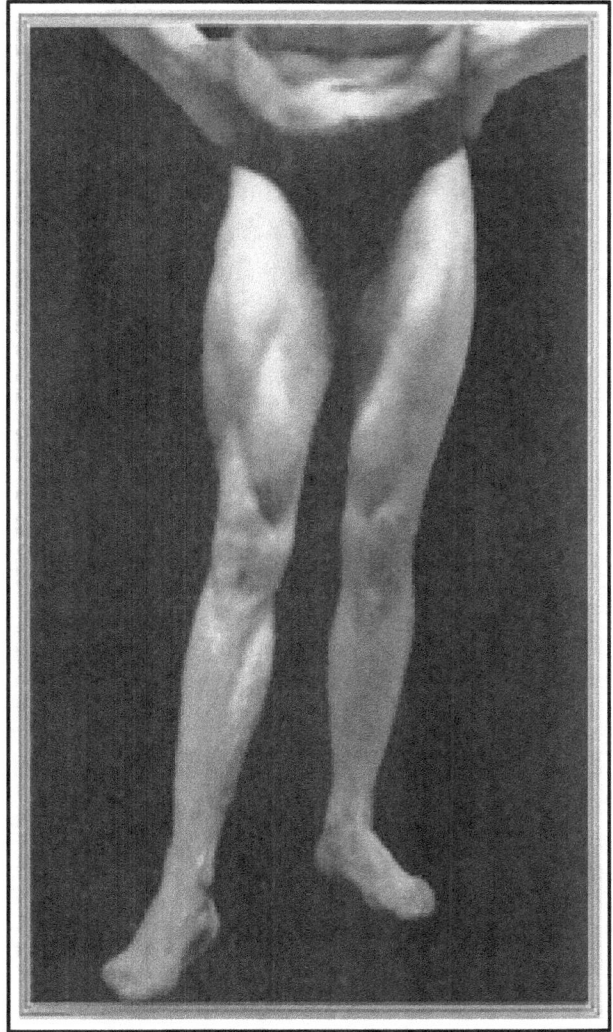

Fig. 49. Bring the extensor muscles into high relief.

CONTRACTION OF THE EXTENSOR MUSCLES OF THE THIGH.

Hang well forward from below the waist, forcing one hip well to the front. (See Fig. 50.)

The torso must bend back to secure balance.

Allow full weight of the whole body to be supported entirely by the front and outside of the thigh.

Fig. 50. Hang well forward from below the waist.

EXERCISE 21.

CONTROL OF THE BICEPS OF THE THIGH, AND GASTROCNEMIUS OF THE CALF.

Secure position shown in Fig. 51.

Point the toes to secure control of the gastrocnemius. Now bring the toes as far to the front as they will come by ankle movement alone, to secure control of the muscles on the outside of the shin. (Tibialis anticus.)

CONTROL OF BICEPS OF THIGH.

Regain position 51, draw the heel of the foot up towards the buttock, concentrating the mind upon the biceps of the thigh.

To secure a more powerful contraction of the biceps of the thigh, the thigh should be carried as far back as it will go.

In securing this control of the biceps of the thigh, the student must go slowly; for cramp is usually experienced in the early stages.

Fig. 50. Initial pose for control of biceps of thigh and calf.

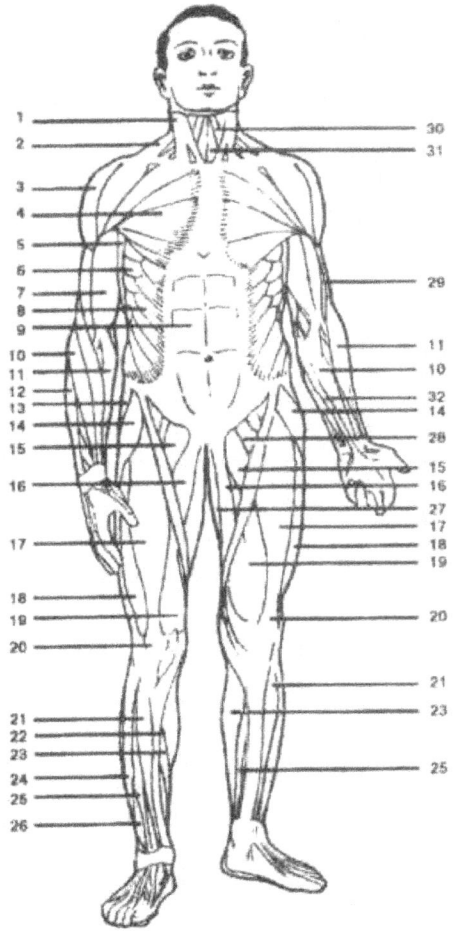

1.	Sterno Cleido Mastoid	12.	Extensor Communis	22.	Gastrocnemius
2.	Trapezius		Digitorum	23.	Soleus
3.	Deltoid	13.	Gluteus Medius	24.	Peroneus Longus
4.	Pectoralis Major	14.	Tensor Vaginae Lemoris	25.	Extensor Proprius Pollicis
5.	Latissimus Dorsi	15.	Pectineous	26.	Peroneus Tertius
6.	Serratus Magnus	16.	Adductor Longus	27.	Gracilis
7.	Biceps	17.	Rectus Femoris	28.	Psoas
8.	Intercostates Externi	18.	Vastus Externus	29.	Brachialis Anticus
9.	Rectus Abdomenus	19.	Vastus Internus	30.	Sterno Hyoid
10.	Supinator Longus	20.	Patella	31.	Omo Hyoid
11.	Flexor Carpi Radialis	21.	Tibialis Anticus	32.	Palmaris Longus

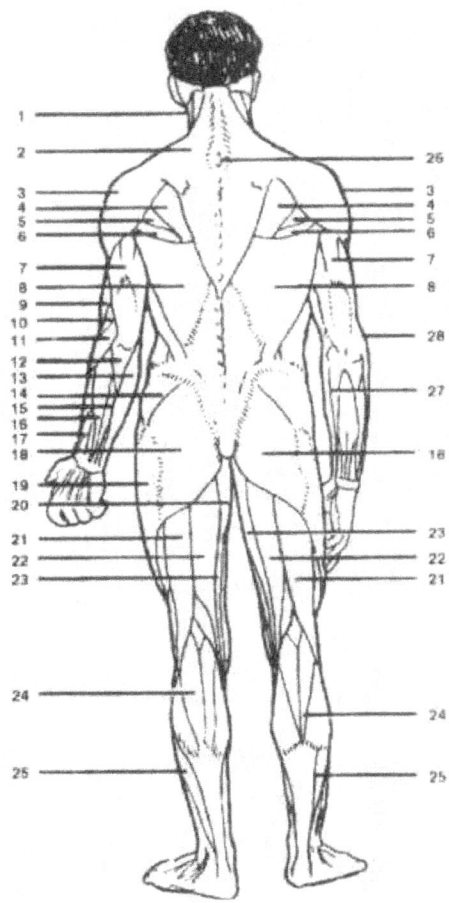

1	Sterno Cleido Mastoid	12	Anconeus	20	Adductor Magnus
2	Trapezius	13	Flexor Carpi Ulnaris	21	Biceps
3	Deltoid	14	Gluteus Medius	22	Semi-Tendinosus
4	Infraspinatus	15	Extensor Carpi Ulnaris	23	Semi-Membranosus
5	Teres Minor	16	Extensor Communis	24	Gastrocnemius
6	Teres Major		Digitorum	25	Soleus
7	Triceps	17	Extensors Ossis Metacarpi	26	7th Cervical Vertebrae
8	Latissimus Dorsi		Pollicis and Primi Internodi	27	Palmaris Longus
9	Brachialis Anticus		Pollicis	28	Olecranon
10	Supinator Longus	18	Gluteus Maximus		
11	Extensor Carpi Radialis	19	Fascia Lata		

For more old time classics of strength, visit:

STRONGMANBOOKS.COM

Currently Available:

The Development of Physical Power by Arthur Saxon

Text Book of Weight Lifting by Arthur Saxon

Bob Hoffman's Simplified System of Barbell Training

Muscle Control by Maxick

More coming soon...

www.ingramcontent.com/pod-product-compliance
Lightning Source LLC
Chambersburg PA
CBHW072142280526
45788CB00002B/741